www.bayegi.com

Parviz Bayegi

With over 15 years of experience in the pharmaceutical industry and successful implementation of numerous GMP- projects, the author brings extensive knowledge and experience to this book.

Email: contact@bayegi.com

PARVIZ BAYEGI

GMP-/FDA calibration & documents

For Pharma/Laboratory/Biotech/ATMP/Medical Device

Parviz Bayegi

Copyright © 2024 Parviz Bayegi

Publishing label:  Bayegi GMP/GXP Consulting

ISBN: 9798320259147

The work, including its parts, is protected by copyright. The author is responsible for the content. Any use is not permitted without authorization.

Table of contents

1. What does calibration mean?.................. 7
2. Standards for calibration 10
3. Calibration & adjustment 11
4. Calibration for GMP qualification........... 13
5. Calibration Types 18
6. Calibration interval 22
7. Accreditation 23
8. Calibration within GMP environment..... 31
9. Traceability... 39
10. In-house or off-site?............................ 40
11. Calibration in Pharma and SOP.......... 43
12. Devices within GMP ranges............... 44
13. Abbreviations..................................... 48
14. Sources 49

Preface

This book provides extensive information about GMP calibrations according to EU-GMP/FDA guidelines for the pharmaceutical industry and laboratories.

The book is suitable for personnel in the field of QA, QC pharmaceutical production and chemical laboratories as well as for students and trainees from the fields of chemistry, pharmaceutical industry, and pharmaceutical engineering.

Have fun and congratulations to the reader.

Your Parviz Bayegi

1. What does calibration mean?

When we talk about calibration, we automatically think of laboratory equipment such as balances or pH-meters. Calibration is a comparison process in which the measured values from the tester are compared and logged with a reference point or reference measuring instrument. If the measured values are within the specified tolerances, the device is called calibrated.

The best example of calibration is the precision scale. Test weights (calibrated) are used as a reference. Various test weights are placed on the precision scale and the respective weight display is logged on the scale. At the end of the logging, the values are analyzed and evaluated. If these values are within the specified tolerance, the precision scale is called calibrated.

To ensure quality assurance, the calibrations are repeated at a specified interval/range.

The references used in the calibrations must be very accurate and have a traceable calibration certificate. Calibrations are necessary wherever the accuracy of the measurements is of great importance.

The measurements are divided into two types:

- One-time measurement

- Permanent measurement

One-time measurement

Figure 1: laboratory scale

The one-time measurement describes the measurement of measuring devices, that are used for one-time measurements and in which the measurement results are directly responsible for the results. This includes measuring instruments such as pH-meters, scales and temperature measuring instruments.

Permanent measurement

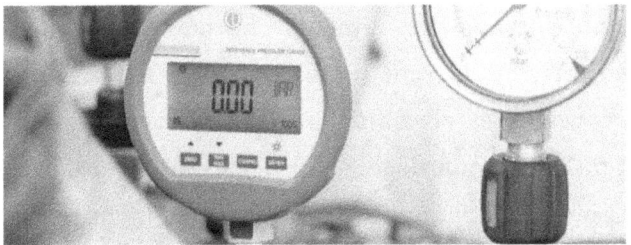

The continuous or permanent measurement describes the measurement of measuring devices that are an integral part of a plant or machine. The correct functioning of such (measuring) devices alone is not sufficient since these devices transmit their measured values to other devices and are processed there. Depending on the measured value, certain processes are started. Examples are the calibration of temperature, pressure and humidity on measuring instruments installed in a WFI system (water for injection production). Without correct readings, this plant cannot obtain the required amount of water.

🔔 By carrying out calibrations, trust in the measuring instruments and systems is built. The calibration will facilitate auditor checks.

---

2. Standards for calibration

There are certain ISO-standards for many industries. Since many different instruments with different properties are suitable for calibration, uniform calibration standards can hardly exist. In order to ensure compliance with the high-quality requirements during calibration, accreditation organizations have been established all over the world. In Germany there is the DAkkS (Deutsche Akkreditierungsstelle GmbH).

One of the most important features of a calibration company is accreditation. The calibrations are very often carried out by a calibration service provider accredited by a national or international accreditation body. In Germany there are many companies that carry out calibrations, such as TÜV Süd, TÜV Nord, Testo, KERN & SOHN GmbH. If a company has been accredited by an accreditation body such as DAkkS (Deutsche Akkreditierungsstelle GmbH), you can trust this calibration company without hesitation.

ISO/IEC 17025

ISO/IEC 17025 is an international standard for the accreditation of testing and calibration laboratories and is mainly used for the calibration and accreditation of laboratories and labor devices. ISO/IEC 17025 contains requirements for the quality management system as well as technical requirements.

3. Calibration & adjustment

Calibration and adjustment have the common goal of providing the specified accuracy, function and correct measurement values.

Adjust

Sometimes the following happens: The display of your personal scale does not point exactly to zero. What now? You manually set the hands as accurately as possible to zero. This type of setting is called adjustment. In the case of an adjustment, intervention on the measuring instrument is required and no reference instrument is required like in calibrations.

Calibration

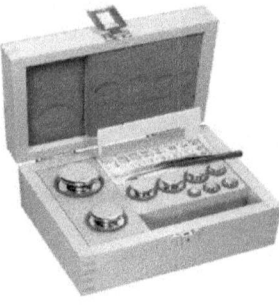

Figure 2: Test weights for scales with calibrated weights

You have probably noticed that your watch does not show the time exactly. You search on the internet for a reference, namely the atomic clock, which displays the International Atomic Time TAI. Now you compare the time from your watch with the atomic clock and then set the time from your watch after the atomic clock. With this you have now calibrated your watch.

In contrast to an adjustment, a very accurately functioning reference is used here for comparison.

Reference standard & calibrated objects

Each calibration consists of a device or object which will be calibrated and the reference standard or reference measuring instrument. The reference standard is very accurate in relation to the speculated object and should be traceable calibrated. Before a device is based on a reference measuring instrument, the reference measuring instrument (reference standard) itself must be calibrated.

Calibrator

With the help of a calibrator, the devices can be calibrated very quickly and easily. In addition to the reference standard, the calibrator sits menu options/software with various functions with various connections and cables. These enable the user to calibrate efficiently.

4. Calibration for GMP qualification

A calibrated device illustrates the accuracy and quality of the expected result. Whether simple measurements or complex processes, calibrations are indispensable in the pharmaceutical industry.

Non-calibration or improper calibration are often the cause of injuries, deaths and even major environmental disasters.

In the case of pharmaceutical products, production costs must be maintained while maintaining high quality and increasing GMP requirements (traceability). In order to comply with these requirements reproducibly for products in the pharmaceutical industry, manufacturers are obliged to implement the GMP guidelines decisively.

An important component of the EU-GMP guideline is the EU-GMP-Annex 15 for the validation/qualification of processes and plants. Due to the EU-GMP-Annex 15 guidelines, all machines, plants, software and processes used for the production of pharmaceutical products must go through an validated and qualified process.

Many of these machines and plants, which have been harmonized within the framework of the EU-GMP Annex 15, include measuring instruments. These measuring instruments must be compatible before the qualification can continue.

Before Validation/Qualification (IQ) can be performed, measuring instruments such as pressure gauges, temperature meters and pH-meters must be calibrated. The calibration certificates must be available during the qualification (FAT/SAT, IQ). These calibration certificates are documented as a proof and part of the qualification.

Figure 3: Structure of a validation process

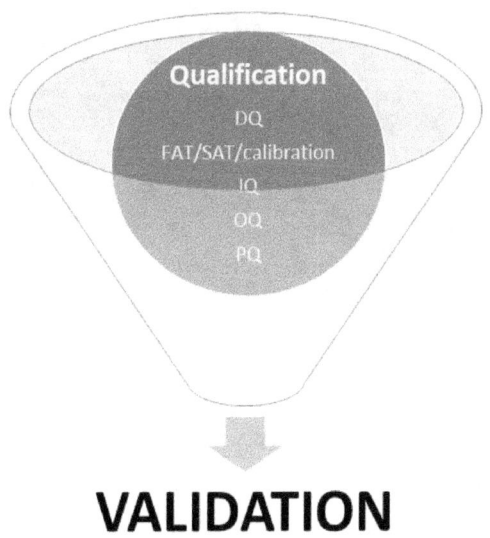

VALIDATION

Example:

Pressure measuring instruments, temperature measuring instruments and pH measuring instruments are integrated into a double-walled stainless-steel tank for mixing raw materials. All these measuring instruments must be calibrated prior to GMP qualification, and the corresponding evidence must be provided.

Figure. 4: Double-walled containers

The necessary safety and quality in the production of pharmaceutical products depend on accurately performed calibrations. Numerous processes in pharmaceutical productions such as CIP/SIP

(Cleaning-In-Place/Sterilization-In-Place) clean systems are also dependent on calibrations.

Numerous products and processes from the industry, and in particular the production of pharmaceutical products such as medicines, have become safer. If you take a medication today without hesitation, you can be sure that this safety has been achieved by constant calibrations of the devices and measuring instruments involved in the product.

5. Calibration Types

The allocated objects /objects and their properties are very different. Considering the complexity of the objects, the calibrations can mainly be divided into simple calibrations and complex calibrations.

Easy calibrations

Simple calibrations are very often calibrations by the comparison with a known value.

Figure 5: Precision scale and calibrated weights

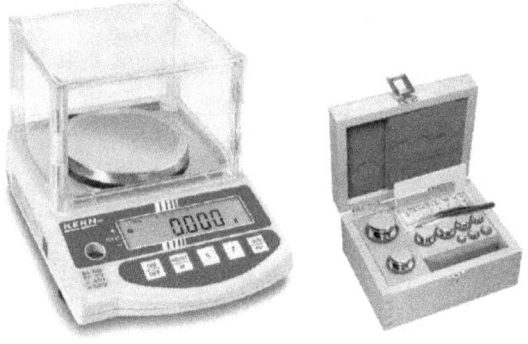

With these calibrations, the results are very quickly visible and the measurements are simple, such as the calibration of laboratory balances. In this calibration, the previously calibrated weight is weighed and the measured value is then read from the scale and logged. If the measured value of the scale with the calibrated weight is equal to or within a specified tolerance, the balance is calibrated. Other devices are pH meters and humidity devices. Such calibrations can also be carried out in-house by trained personnel such as laboratory staff.

Complex calibrations

In complex calibrations, the calibration of the instruments is more complex and up to five measurements and comparisons with reference instruments can take place in addition to simple calibrations.

Figure 6: Professional data logger

The surrounding properties such as temperature, humidity and electromagnetic influences are also very important for complex calibrations. Devices for complex calibrations are for example particle counters, devices with high-frequency technology, HPLC and measuring instruments from industry and research.

Figure 7: Handy particle counters

Figure 8: Sample calibration certificate for thermometers

KERN & SOHN GmbH

CALIBRATION

Kalibrierlabor seit 1994.
Calibration laboratory since 1994.

Ihr Partner für Kalibrierdienstleistungen, Prüfmittelmanagement und Beratung.
Your partner for calibration services, test equipment management and support.

Kalibrierschein *Calibration Certificate*	**Muster-KERN-2020-12**	
Kalibriergegenstand *Calibration object*	Elektronisches, digitales Thermometer *Electronic, digital thermometer*	Dieser Kalibrierschein dokumentiert die bestimmungsgemäße Messfunktionalität des Kalibriergegenstands, die sich in Einheiten des Internationalen Einheitensystems (SI) ausdrückt und unter Zuhilfenahme von Messhilfsmitteln ermittelt wurde, die sich auf entsprechende nationale Normale zurückführen lassen.
	Messbereich: -10 °C bis 60 °C *Range*	
	Ablesbarkeit: 0,1 °C *Accuracy*	
Hersteller *Manufacturer*	Dostmann/Wertheim	
		Für die Einhaltung einer angemessenen Frist zur Wiederholung der Kalibrierung ist der Benutzer verantwortlich.
Typ *Type*	30.2017.02	
Fabrikat/Serien-Nr. *Serial number*	K18123456	*This calibration certificate documents the intended function of measurement of the calibrated object which is expressed in units of the "Le Système internation d'unités" (SI). The measurement was executed with the aid of measurement utilities which are traceable to national standards.*
Inventar-Nr. *Inventory number*	-	
Auftraggeber *Customer*	Mustermann GmbH Musterweg 42 12345 Musterstadt Deutschland	*The user is obliged to have the object recalibrated at appropriate intervals.*
		Die englische Übersetzung des Kalibrierscheines ist eine unverbindliche Übersetzung. Im Zweifelsfall gilt der deutsche Originaltext.
Auftragsnummer *Order No.*	2020-123456789	
Datum der Kalibrierung *Date of calibration*	16.12.2020	*The English version of the calibration certificate is not a binding translation. If any matters give rise to controversy, the German original text must be used*
Ort der Kalibrierung: *Place of calibration*	Labor 8 - Platz 1 *Calibration laboratory KERN*	
Kalibrierverfahren: *Calibration method*	Der Kalibriergegenstand und das Normal werden zur Temperaturangleichung im klimatisierten Raum in unmittelbarer Nähe zueinander platziert. Nach einer Zeit von 24 Stunden wird das Thermometer mit dem Bezugsnormal verglichen. *The instrument and the standard were placed in an air-conditioned room for acclimation. After 24 hours the thermometer are compared to the standard.*	
Bezugsnormal: *Reference standard*	U_TR_1	Bemerkungen: - *Remarks*

Messergebnisse / *Measurement results*

Anzeige / *Indication*		Messabweichung	erw. Unsicherheit	
Normal *Standard*	Prüfling *Cal. object*	*Error*	*exp. uncertainty*	k = 2 (95%)
21,87 °C	21,6 °C	-0,27 K	± 0,5 K	

Dieser Kalibrierschein darf nur vollständig und unverändert weiterverbreitet werden. Auszüge oder Änderungen bedürfen der Genehmigung des ausstellenden Kalibrierlaboratoriums. Kalibrierscheine ohne Unterschrift haben keine Gültigkeit.

This calibration certificate may not be reproduced other than in full except with the permission of the issuing laboratory. Calibration certificates without signature are not valid

| KERN & SOHN GmbH Postfach 4052 72322 Balingen-Frommern Tel.: 07433 - 99 33-0 Fax: 07433 - 99 33-149 E-mail: info@kern-sohn.com | Datum *Date* 16.12.2020 | Leiter des Kalibrierlaboriums *Head of the calibration laboratory* Otto Grunenberg | Bearbeiter *Person in charge* Roswitha Komrowski |

KERN & Sohn GmbH, Ziegelei 1, 72336 Balingen, Germany
Phone +49-[0]7433-9933-0, Fax +49-[0]7433-9933-149

> 🔔 In both methods, a calibration certificate or verification of the reference instrument must be available prior to the calibration of instruments /objects (see figure 8).

6. Calibration interval

The adherence to intervals for calibrations is very important, because a complete calibration interval ensures the reproducible measurement and accuracy of the device.

Failures to observe predetermined intervals can affect the accuracy of the measuring instrument, which can cause high damages.

The appropriate intervals are determined by the manufacturers. Very often an annual calibration interval is given. If a measuring instrument is used very often, deviations or inaccuracies may occur before the planned recalibration. In such cases, the interval can be shortened.

One example is the laboratory scale. For laboratory scales, a calibration interval of at least 12 months is given, depending on the precision. However, if a laboratory balance is used very often and daily, this calibration interval can be shortened to 6 months.

7. Accreditation

In order to increase confidence in a company offering a calibration service, it is necessary that a third party organization (national or international) audits these companies for the calibration service and confirms that the company is responsible for the calibration service and carries out the calibrations correctly. This process of auditing by third parties is called accreditation. Subsequently, a calibration service provider can call itself an accredited company (see illustration below/ example calibration certificate from company company).

Each country has its own and sometimes several accreditation organizations (see table below). These national organizations are responsible for the accreditation of national and international companies. This means that a company in China can obtain accreditation from a German accreditation organization DAkkS or ANSI in the USA because of international business.

International Accreditation Organizations

In order to recognize and harmonize national guidelines for calibrations between countries, some national organizations have become members of international organizations such as ILAC (International Organization for Accreditation Body) and IAF (International Accreditation Forum).

ILAC

Most national accreditation bodies are members of ILAC. ILAC is an international association of accreditation bodies and operates in the field of ISO/IEC 17011 and in the accreditation of conformity assessment bodies, including calibration laboratories (ISO/IEC 17025), testing laboratories (ISO/IEC 17025), medical laboratories (ISO 15189) and inspection bodies (ISO/ IEC 17020).

> 🔔 ILAC and the International Accreditation Forum (IAF) have been working together since 2001. The aim is to create a single international accreditation organization.

ILAC manages these international agreements in the field of accreditation of calibration laboratories, testing laboratories, medical laboratories and inspection bodies and IAF in the areas of management system certification, product certification, services, personal certification and other similar conformity assessment programs. Both organizations, ILAC and IAF, work together and coordinate their efforts to improve accreditation and conformity assessment worldwide.

source: www.ilac.org

IAF

The International Accreditation Forum (IAF) is a worldwide association of accreditation bodies and other bodiesinterested in conformity assessment in the areas of management systems, products, processes, services, personnel, validation and verification and other similar conformity assessment programs.

Our primary mission is to develop a single global conformity assessment program that reduces risk to companies and their customers by assuring them that accredited certificates and validation and verification declarations can be reliable.

source: www. Iaf.org

Benefits of Membership of International Organizations

The international agreements ensure that national accreditation organizations are recognized worldwide. Thus, by the membership of the calibration companies in an international organization such as ILAC/IAF, the calibration can be easily accepted in another country.

Table: National accreditation organizations

Deutsche Akkreditierungsstelle GmbH – **DAkkS**	Germany
National Standardization Council of Thailand - **NSC**	Thailand
French Accreditation Committee - **COFRAC**	France
ENAC	Spain
Czech Accreditation Institute - **CAI**	Czech Republic
TRIBUTE	Denmark
Federal Ministry of Digital and Economic Affairs	Austria
Belgian Organization for Accreditation - **BELAC**	Belgium
Hellenic Accreditation System S.A. - **ESYD**	Greece

National Accreditation Centre of Iran **NECI**	Iran
Irish National Accreditation Board - **INAB**	Ireland
National Accreditation System - **ACCREDIA**	Italy
Dutch Accreditation Council - **RvA**	Netherlands
Norwegian Accreditation - **NA**	Norway
IPAC	Portugal
Swedish Board for Accred. & Conformity Assessment - **SWEDAC**	Sweden
Swiss Accreditation Service - **SAS**	Switzerland
Turkish Accreditation Agency - **TÜRKAK**	Turkey
The Finnish Accreditation Service - **FINAS**	Finland
Slovenian Accreditation- **SA**	Slowenia
Polish Centre for Accreditation - **PCA**	Poland
Argentine Accreditation Agency - **OAA**	Argentina
Korea Accreditation System - **KAS**	Korea
CNAS	China
THEM	Australia
ANZ	New Zealand
INMETRO	Brazil
Entidad Mexicana de Acreditacion, a.c. - **EMA**	Mexico

South Africa National Accreditation System - **SANAS**	South Africa
United Kingdom Accreditation Service - **UKAS**	United Kingdom
American National Standards Institute - **ANSI**	United States of America
National Accreditation Board for Certification Bodies **(NABCB)**	India
Japan Accreditation Board - **JAB**	Japan

Calibration service providers should be certified by a national or international accreditation organization. Only then is it possible to be sure that the calibrations comply with the required standardized standards and qualities.

Some of these standards are:

- Testing ISO/IEC 17025 & ISO 15189
- Calibration ISO/IEC 17025
- Inspection ISO/IEC 17020
- Proficiency Testing Providers ISO/IEC 17043
- Reference Materials Producers ISO 17034

Whether a calibration service provider has been internationally or internationally accredited can be determined by the corresponding declarations on the respective calibration certificate (see image section below). The homepage of the calibration provider can also inform about the accreditation.

Figure 9: Detail from the example calibration certificate with accreditation Design by DAkkS/DKB/iLac from Kern & Sohn GmbH-Germany

> 🔔 DKD In accordance with Regulation (EC) No. 765/2008, the accreditation system in Germany was changed on 1st January 2010. The accreditation body of the German Calibration Service (DKD) was transferred to the Deutsche Akkreditierungsstelle GmbH (DAkkS) with effect from 17.12.2009.
>
> source: www.dkd.eu

8. Calibration within GMP environment

Calibration certificates may vary by country and industry standard.

A calibration certificate, which was created, for example, for a scale in a food trade, is very simple in view of its tasks. In contrast, a calibration certificate created for a precision laboratory scale (within GMP environment) has special requirements and content.

The calibration certificates are very important documents for any GMP (validation/qualification) and

GMP/FDI audit. Improper documentation can be a problem during the tests and lead to complaints.

Based on the sample calibration shown below (Figure 10), I would like to give you an impression of a professionally created calibration certificate/proof.

Enterprise

A calibration certificate must contain sufficient information about the company performing the calibration.

Which accreditation organization?

Sometimes a calibration certificate contains only the sentence "... the company is accredited... ". This statement is not sufficient or trustworthy. It is essential that the exact name and logo of the accreditation organization be displayed in a calibration certificate. If a company claims to be accredited then they should provide their membership number.

Calibration number

Next, a number and date should be assigned to the calibration.

Object

Part of a calibration is also the information about the object which is calibrated. In our example it is a weight set with a margin of 1 mg-1 kg from the company Kern & Sohn GmbH.

Ambient conditions

A calibration should be carried out in accordance with certain technical specifications and values. In our example below, the weights (approaches) are sensitive to temperature and humidity. Therefore, the environment temperature and humidity should be reported during the calibration.

Reference standards

The extreme measurement accuracy during calibration is achieved by comparing measured values with a reference device (reference standards). The following information about the reference instrument used shall be entered in a calibration certificate:

- Name and serial number of the reference devices

- Calibration of the reference device

- Validity date of the calibration of the reference device

- If necessary, the corresponding tolerances

If the measured values are identical to the reference value or within the tolerance, the results are documented and drawn as calibrated. After that, the object can be described as calibrated.

How many test series?

In order to exclude errors in measurements or to get an average value, the tests are carried out more than once. Many tests are repeated three to five times. How many test series are necessary to successfully perform a calibration depends on the environment constellations and tasks or complexity of the instrument which is calibrated.

If, for example, the instrument to be calibrated is located and is intended to function in a constantly changing environment temperature and humidity, then the tests are repeated several times in different ambient temperatures and humidities.

The tasks of the device are another important reason to carry out tests more than once. For example, if the device to be calibrated is part of a very important quality process (for example, a particle counter in a GMP Class A isolator), then the test is carried out more than three times.

Figure 10: Calibration of weight set with accreditation

KERN & SOHN GmbH

Akkreditiertes Kalibrierlabor seit 1994.
Accredited calibration laboratory since 1994

Ihr Partner für Kalibrierdienstleistungen, Prüfmittelmanagement und Beratung.
Your partner for calibration services, test equipment management and support

Mitglied im / *member of the*

Deutschen Kalibrierdienst

	Sample
	D-K 19408-01-00

| Kalibrierschein *Calibration certificate* | Sample-2020-01/1 | Kalibrierzeichen *Calibration mark* | 2020-01 |

Gegenstand *Object*	Gewichtssatz, 1 mg - 1 kg Klasse E2 *Set of weights, 1 mg - 1 kg Class E2*	Dieser Kalibrierschein dokumentiert die Rückführung auf nationale Normale zur Darstellung der Einheiten in Übereinstimmung mit dem Internationalen Einheitensystem (SI).
Hersteller *Manufacturer*	KERN & Sohn GmbH Ziegelei 1 D-72336 Balingen Germany	Die DAkkS ist Unterzeichner der multilateralen Übereinkommen der European co-operation for Accreditation (EA) und der International Laboratory Accreditation Cooperation (ILAC) zur gegenseitigen Anerkennung der Kalibrierscheine.
Typ *Type*	313-052	Für die Einhaltung einer angemessenen Frist zur Wiederholung der Kalibrierung ist der Benutzer verantwortlich.
Fabrikate/Serien-Nr. *Serial number*	G123456789	
Auftraggeber *Customer*	Mustermann GmbH	*This calibration certificate documents the traceability to national standards, which realize the units of measurement according to the International System of Units (SI). The DAkkS is signatory to the multilateral agreements of the European co-operation for Accreditation (EA) and of the International Laboratory Accreditation Cooperation (ILAC) for the mutual recognition of calibration certificates. The user is obliged to have the object recalibrated at appropriate intervals.*
Auftragsnummer *Order No*	2020-123456789	
Anzahl der Seiten des Kalibrierscheines *Number of pages of the certificate*	3	
Datum der Kalibrierung *Date of calibration*	10.01.2020	

Dieser Kalibrierschein darf nur vollständig und unverändert weiterverbreitet werden. Auszüge oder Änderungen bedürfen der Genehmigung des ausstellenden Kalibrierlaboratoriums. Kalibrierscheine ohne Unterschrift haben keine Gültigkeit.
This calibration certificate may not be reproduced other than in full except with the permission of the issuing laboratory. Calibration certificates without signature are not valid.

Datum *Date*	Leiter des Kalibrierlaboratoriums *Head of the calibration laboratory*	Freigabe des Kalibrierscheins durch *Approval of the calibration certificate by*
16.01.2020	Grunenberg	Rocco Scaramuzzo

KERN & SOHN GmbH, Ziegelei 1, D-72336 Balingen, Germany
Phone +49-7433-99330, Fax +49-7433-9933-149

Seite 3 zum Kalibrierschein vom 16.01.2020
Page 3 of the calibration certificate dated

Sample

D-K
19408-01-00

2020-01

Messergebnisse:
Measurement results:

Nennwert	Kennzeichnung	konventioneller Wägewert	Unsicherheit $k=2$	Fehlergrenze	Klasse*
nominal value	marking	conventional mass	uncertainty	max. perm. error	class*
1 mg		1 mg + 0,0010 mg	0,0020 mg	± 0,0060 mg	E2 ✓
2 mg		2 mg + 0,0005 mg	0,0020 mg	± 0,0060 mg	E2 ✓
2 mg	*	2 mg + 0,0016 mg	0,0020 mg	± 0,0060 mg	E2 ✓
5 mg		5 mg + 0,0010 mg	0,0020 mg	± 0,0060 mg	E2 ✓
10 mg		10 mg + 0,0009 mg	0,0020 mg	± 0,0080 mg	E2 ✓
20 mg		20 mg - 0,001 mg	0,003 mg	± 0,010 mg	E2 ✓
20 mg	*	20 mg + 0,001 mg	0,003 mg	± 0,010 mg	E2 ✓
50 mg		50 mg + 0,001 mg	0,004 mg	± 0,012 mg	E2 ✓
100 mg		100 mg + 0,001 mg	0,005 mg	± 0,016 mg	E2 ✓
200 mg		200 mg + 0,002 mg	0,006 mg	± 0,020 mg	E2 ✓
200 mg	*	200 mg + 0,003 mg	0,006 mg	± 0,020 mg	E2 ✓
500 mg		500 mg + 0,005 mg	0,008 mg	± 0,025 mg	E2 ✓
1 g		1 g + 0,002 mg	0,010 mg	± 0,030 mg	E2 ✓
2 g		2 g + 0,002 mg	0,013 mg	± 0,040 mg	E2 ✓
2 g	*	2 g + 0,002 mg	0,013 mg	± 0,040 mg	E2 ✓
5 g		5 g + 0,010 mg	0,016 mg	± 0,050 mg	E2 ✓
10 g		10 g - 0,007 mg	0,020 mg	± 0,060 mg	E2 ✓
20 g		20 g + 0,005 mg	0,026 mg	± 0,080 mg	E2 ✓
20 g	*	20 g + 0,015 mg	0,026 mg	± 0,080 mg	E2 ✓
50 g		50 g + 0,02 mg	0,03 mg	± 0,10 mg	E2 ✓
100 g		100 g + 0,01 mg	0,05 mg	± 0,16 mg	E2 ✓
200 g		200 g + 0,05 mg	0,10 mg	± 0,30 mg	E2 ✓
200 g	*	200 g - 0,00 mg	0,10 mg	± 0,30 mg	E2 ✓
500 g		500 g + 0,10 mg	0,26 mg	± 0,80 mg	E2 ✓
1 kg		1 kg + 0,1 mg	0,5 mg	± 1,6 mg	E2 ✓

* Bewertung der Klasse bzw. der Fehlergrenze (wenn keine Klassenangabe vorhanden ist) bezieht sich nur auf den konventionellen Wägewert.
The assessment of the class / the max. perm. error (if no class assessment is given) only refers to the conventional mass

Angegeben ist die erweiterte Messunsicherheit, die sich aus der Standardunsicherheit durch Multiplikation mit dem Erweiterungsfaktor $k=2$ ergibt. Sie wurde gemäß EA-4/02 M 2013 ermittelt. Der Wert der Messgröße liegt mit einer Wahrscheinlichkeit von 95% im zugeordneten Werteintervall.
Die erweiterte Messunsicherheit wurde aus Unsicherheitsanteilen der verwendeten Normale, der Wägungen und der Luftauftriebskorrektur berechnet. Eine Abschätzung über Langzeitveränderungen ist in der Unsicherheitsangabe nicht enthalten.
Reported is the expanded uncertainty which results from the standard uncertainty which results from the standard uncertainty by multiplication with the coverage factor $k=2$. It has been evaluated according to EA-4/02 M 2013.
The value of the measurand is found within the attributed interval with a probability of 95%.
The expanded uncertainty was calculated from the contributions of uncertainty originating from the standards used, from the weighings and the air buoyancy corrections. The reported uncertainty does not include an estimate of long-term variations.

Bemerkungen: Das Kalibrierlaboratorium bewahrt eine Kopie dieses Kalibrierscheins für mindestens 5 Jahre auf.
Remarks: *The calibration laboratory retains a copy of this calibration certificate for at least 5 years.*

Ende des Kalibrierscheines
End of calibration certificate

KERN & SOHN GmbH, Ziegelei 1, D-72336 Balingen, Germany Sec: [1da70]
Phone +49-7433-99330, Fax +49-7433-9933-149   MXCG101EN (rev 5)

Seite 2 zum Kalibrierschein vom 16.01.2020
Page 2 of the calibration certificate dated

Sample
D-K-19408-01-00
2020-01

Die englische Übersetzung des Kalibrierscheines ist eine unverbindliche Übersetzung.
Im Zweifelsfall gilt der deutsche Originaltext.
The English version of the calibration certificate is not a binding translation.
If any matters give rise to controversy, the German original text must be used.

Kalibriergegenstand: Gewichtssatz, 1 mg - 1 kg
Calibration object Klasse E2

Set of weights, 1 mg - 1 kg
Class E2

Untergebracht in einem Etui.
Located in a box.

Kalibrierverfahren: Die Kalibrierung erfolgte durch Vergleich mit den Bezugsnormalen
Calibration method des Kalibrierlaboratoriums nach der Substitutionsmethode mit Auftriebskorrektur.
The calibration ensued through comparison with the reference standards of the
calibration laboratory using the substitution method with air buoyancy correction.

Ort der Kalibrierung: Kalibrierlaboratorium KERN
Place of calibration *Calibration - Laboratory KERN*

Umgebungsbedingungen: Die Kalibrierung wurde bei folgenden Umgebungsbedingungen ausgeführt:
Ambient conditions *The calibration was carried out under the following ambient conditions:*

	von *from*	bis *to*	Unsicherheit *uncertainty*
Temperatur (°C) *temperature*	22,9	24,1	0,1
rel. Luftfeuchte (%) *relative humidity*	48,5	53,4	2,0
Luftdruck (hPa) *air pressure*	942,5	948,5	0,3

Magnetische Eigenschaften: Der Hersteller hat bestätigt, dass die Gewichtsstücke die magnetischen
Magnetic properties Eigenschaften gemäß R111:2004 einhalten.
The manufacturer has confirmed the compliance of the magnetic properties of the weight pieces with the
OIML R111:2004.

Referenzgewichte: 123-D-K-19408-01-00-2019-05
Standard weights

Material / angenommene Dichte:
Material / assumed density

Nennwert *nominal value*	Dichte *density*	Unsicherheit *uncertainty*	Material *material*	Form *shape*
1 mg - 500 mg	7950 kg/m^3	140 kg/m^3	Edelstahl *Stainless steel*	Draht *Wire*
1 g - 1 kg	8000 kg/m^3	100 kg/m^3	Edelstahl *Stainless steel*	Knopf *Cylindrical form*

KERN & SOHN GmbH, Ziegelei 1, D-72336 Balingen, Germany Sec: [1da70]
Phone +49-7433-99330, Fax +49-7433-9933-149 MXCG101EN (rev 5)

9. Traceability

For calibrations in pharmaceutical and medical technology (in GMP environments) as well as in aerospace technology, the traceability of calibration is of enormous importance. You have read above about the importance of reference devices and now know how important a calibrated reference device is. Traceability means that the equipment which is used (e.g. measuring instruments such as particle counters and radiation measuring instruments) has been calibrated continuously and is also verifiably documented.

Only through complete traceability of the calibration verifications can a reproducible quality and accuracy of the product be guaranteed.

10. In-house or off-site?

In the pharmaceutical industry, depending on availability, the calibrations of devices and measuring instruments are carried out by two types:

- IN-HOUSE calibration
- EXTERNAL calibration

The procedures for both types of calibration, in-house and external, are regulated by an SOP.

IN-HOUSE calibration

In the case of an in-house calibration, the calibration of the devices is carried out in operation by trained personnel. Often the costs and the large number of calibrations are the reason for an in-house calibration. A good example is the calibration of laboratory scales. When calibrating laboratory scales, a company can buy the calibrated weights and after the responsible person has received the necessary training, all laboratory trolleys can be calibrated with the calibrated weights. Even with an in-house calibration, there must exist calibration certificates with corresponding contents (see 8. Calibration certificate in content).

EXTERNAL calibration

The external calibration involves a company tasked with calibrating specific devices and instruments. The decision to commission extern companies is often associated with the fact that the calibrations are very complex and the corresponding measuring instruments for the calibrations are very expensive.

Logbook

It is necessary that a logbook exists for quality-relevant devices. All calibrations and changes are documented in this logbook.

Figure 11: Logbook for mobile particle counters

Logbook for mobile particle countingr

Manufacturer Device Series No. xxxxxxxx Internal device No.: xxxxxxxxxx
Responsible SOP: xxxxxxxxxxx
Date of first installation: xxxxxx
Place of use: xxxxxxx

Event	Date	Carried out by/Company	Documents available	Further explanation
Calibration	01.02.2023	P.Bayegi/QA	YES☐ No ☐	
			YES ☐ No ☐	
			YES ☐ No ☐	

11. Calibration in Pharma and SOP

In the pharmaceutical industry, the requirements for a device or measuring device are particularly high. Therefore, the devices are divided into two groups according to GMP guidelines:

- Non-GMP equipment
- Equipment within GMP ranges

Non-GMP equipment

Devices that are installed outside GMP areas are devices that function outside cleanrooms and therefore have no influence on the GMP-guidelines for the respective production and processes. Such devices would not be qualified, such as measuring instruments used in energy production or water treatment. For these devices, the calibrations are carried out according to maintenance intervals.

12. Devices within GMP ranges

All companies that work with a quality management system (EU-GMP, ISO 9000, PICs, FDA or similar) have a calibration required. Through the quality management system, the company undertakes to meet standard requirements such as ISO/IEC 17025 calibration and to test the measuring instruments at regular intervals. All calibrations must be documented.

In order to meet the high-quality requirements in a GMP cleanroom, all quality-relevant processes such as calibrations including intervals are determined by a SOP procedure (see Figure 12).

Figure 12: SOP for Instrument Calibration (3 page)

Firma : XY Pharma
SOP for instrument calibration

Seite *1 von 3*
SOP-Nr.: *XXXXXX*
Version: *1.0*
Datum: *01.01.2023*

SOP for instrument calibration (internal and third-party)

1.0 Document for submission

Tasks	Name	Function	Date, signature
Creator			
Tested by			
Approval by			

2.0 OBJECTIVE

The purpose of this SOP is to describe the procedure for the calibration plan and calibration practices of measuring instruments and equipment.

3.0 SCOPE

This SOP applies to the calibration of measuring instruments, instruments and equipment that are serviced in the pharmaceutical product production facility.

Firma : XY Pharma	**Seite** *2 von 3*
SOP for instrument calibration	**SOP-Nr.:** *XXXXXX*
	Version: *1.0*
	Datum: *01.01.2023*

4.0 RESPONSIBILITY

- Quality Control (QC)
- Quality Security (QA)
- Engineering Department

5.0 PROCEDURE

Instrument calibration and scheduling procedures:

5.1 Preparation of the calibration plan and updating of the instrument/instrument master list

5.2 Quality Control creates a schedule for the following calibrations:

a- Internal
b- External (by third parties)

5.3 The documents in Appendix 1,2,3 are used as templates.

5.4 Upon purchase of a new instrument or equipment, it will be added to the master list and the necessary calibration intervals will be made in accordance with this SOP.

5.5 Depending on the quality risks, a logbook is created. If there are several small instruments within a device/system, a logbook is created for all integrated instruments.

Firma : XY Pharma *SOP for instrument calibration*	**Seite** 3 von 3 **SOP-Nr.:** XXXXXX **Version:** 1.0 **Datum:** 01.01.2023

6.0 ABBREVIATIONS

QA	Quality Assurance
QC	Quality control
SOP	Standard Operating Procedure

7.0 Appendix

Appendix 1: Internal calibration template

Appendix 2: External calibration template

Appendix 3: Template Logbook

13. Abbreviations

CIP	Cleaning-In-Place
FAT	Factory Acceptance Test
FDA	Food and Drug Administration
IQ	Installation Qualification/ Installationsqualifikation
GMP	Good Manufacturing Practice
PICs	Pharmaceutical Inspection Co-operation Scheme
QA	Quality Assurance
QC	Quality control
SIP	Sterilization-In-Place
SOP	Standard Operating Procedure / Standardarbeitsanweisung
VDI	Association of German Engineers
WFI	Water for injections SIP-CIP systems

14.   Sources

- ISO/IEC 17025
- EU-GMP guidelines
- International Association of Accreditation Bodies (ILAC)
- International Accreditation Forum (IAF)
- Company Kern & Sohn GmbH (pictures and samples calibration certificates)
- Trotec GmbH (pictures)

Notiz:

GMP

GMP-/FDA- Reinraumplanung & Pharma-Engineering

für Pharma / Biotech / ATMP / Medical Device

GMP

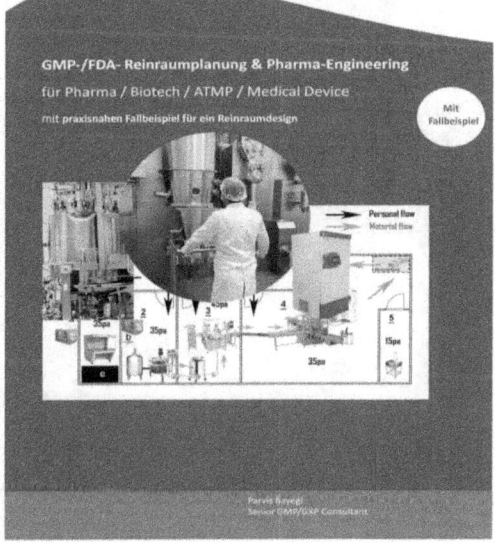

Themen in diesem Buch:

- Reinraum-Standardisierungen
- Richtlinien (GMP, FDA, PICs) für die Produktion im Reinraum
- ISO/Normen für den Reinraumbau
- GMP-Reinraumbau und Design
- Reinraum- Gebäudebestandteile
- Lüftungstechnik im Reinraum
- Reinraum-Monitoring & Messungen
- Werkstoffe & Oberflächen im Reinraum
- Fallstudie für die Planung eines GMP-Reinraums

FMEA fast & professional for GMP/cGMP projects

Step-by-step learning with colored content, templates and case studies

Topics in this book:
- FMEA goals and areas of application
- FMEA process flow
- FMEA types
- When should an FMEA be carried out?
- FMEA step-by-step preparation
- FMEA step-by-step implementation
- Flowchart for FMEA process
- FMEA case studies and templates

CCS (Contamination Control Strategy) EU-GMP-Annex 1/2022

Topics in this book:

- QRM & CCS
- The Contamination Control Strategy (CCS) - What is it?
- How should a CCS umbrella document look like?
- What elements/areas does CCS contain?
- Example/template for CCS-Umbrella-Document
- Example/template for CCS excel list
- Internet Link for Template download

GMP -/FDA- Cleanroom planning & pharmaceutical engineering with practical case studies for

Topics in this book:
- Cleanroom standard/ ISO 14644
- Guidelines (GMP, FDA, PICs) for cleanroom production
- ISO/Standards for cleanroom construction
- GMP cleanroom construction and design
- Cleanroom construction/ components
- Ventilation technology in the cleanroom
- Cleanroom monitoring & measurements
- Materials & surfaces in the cleanroom
- Case study for the design of a GMP cleanroom

GMP Compliance at Validation, Qualification & Docume

with practical case studies and templates for

Pharma / Biotech / ATMP / Medical Device   ISBN   978-1-4478-5510-1

Topics in this book:

What is Qualification and what is Validation? Why do I qualify?

How do I get started with a GMP concept/project?

What are my GMP Qualification strategies?

How do I write a project risk analysis?

What is Change Control (CC) and do I need a Master or Sub CC?

How do I write a Validation Master Plan (VMP)?

What is a FMEA and why do I need a FMEA?

How do I write a FMEA? How do I write a Qualification plan (QP)?

What are FAT & SAT? And do I need these tests?

How do I create Qualification documents (DQ,IQ,OQ,PQ)?

Step by step to Validation and Qualification based on case studies

GMP

EU-GMP- Annex 1-2022
mit deutscher Übersetzung, Tipps und Hinweise

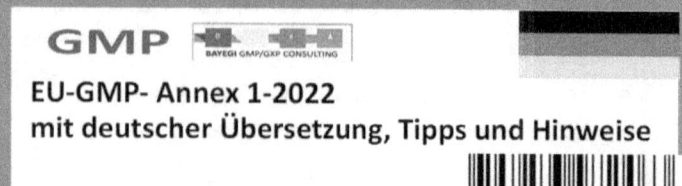

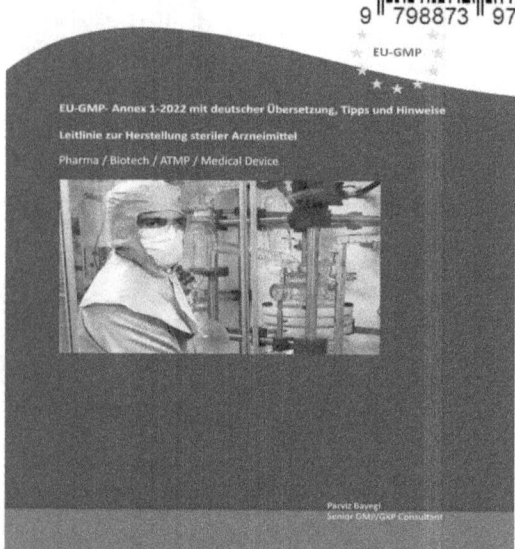

EU-GMP- Annex 1-2022

Leitlinie zur Herstellung steriler Arzneimittel

GMP- gerechte Validierung/Qualifizierung & Dokumentation
Mit praxisnahen Fallbeispielen und Templates für

Pharma / Biotech / ATMP / Medical Device

Themen in diesem Buch:

Was ist Qualifizierung und was ist Validierung?
Warum qualifiziere ich?
Wie beginne ich mit einem GMP-Konzept/Projekt?
Wie lauten meine GMP-Qualifizierungsstrategien?
Wie schreibe ich eine Projektrisikoanalyse?
Was ist Change control (CC) und brauche ich ein Master oder Sub CC?
Wie schreibe ich einen Validierungsmasterplan (VMP)?
Was ist ein FMEA und wozu brauche ich ein FMEA?
Wie schreibe ich ein FMEA?
Wie schreibe ich einen Qualifizierungsplan (QP)?
Was sind FAT & SAT? Und brauche ich diese Tests?
Wie erstelle ich Qualifizierungsdokumente (DQ,IQ,OQ,PQ)?
Schritt für Schritt zur Validierung und Qualifizierung anhand von Fallbeispielen

GMP

GMP- FAT & SAT Konzept und Durchführung

Mit praxisnahen Fallbeispielen und Template

für Pharma / Biotech / ATMP / Medical Device

GMP

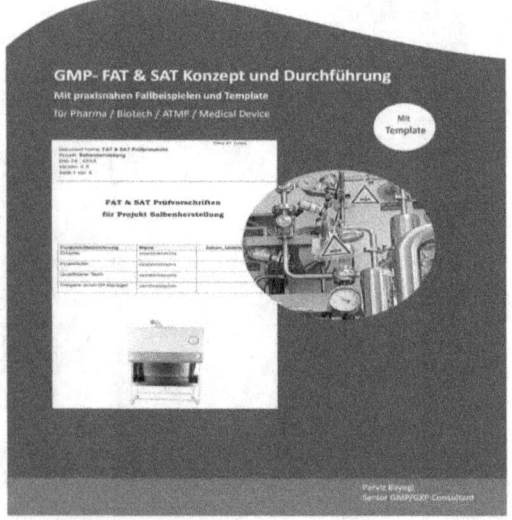

Themen in diesem Buch:

- Warum brauchen wir GMP-FAT/SAT?
- Besonderheiten von FAT & SAT bei GMP-Qualifizierung
- Wichtige FAT/SAT-Tests
- Implementierung in IQ & OQ
- FAT & SAT Konzept erstellen und schreiben
- Aufbau des Dokuments Prüfungsvorschriften
- Fallbeispiel für FAT/SAT-Dokumente
- Fallstudie FAT/SAT Prüfungen und Testprotokolle

GMP

GMP- Lastenheft & Pflichtenheft für die Bestellprozesse

von Geräten & Anlagen innerhalb Reinräume und GMP-Projekten

GMP

Themen in diesem Buch:

- Bestellprozesse & Dokumentierung bei GMP-Reinraumprojekten
- Allgemeine Anforderungen für GDP- & GMP-Dokumente
- GMP-Lastenheft und Qualifizierung
- Vorbereitung für GMP-Lastenheft
- URS
- Bestandteile eines GMP-Lastenheft & Pflichtenheft
- Schreiben eines GMP-Lasterhafts anhand eines Fallbeispiels

GMP- Kalibrierungen & SOP-Kalibrierungsdokumente

für Pharma / Biotech / ATMP / Medical Device

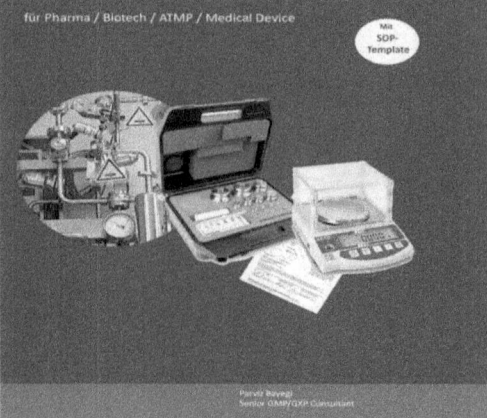

Themen in diesem Buch:

-Kalibrierung in Pharma/Labor

-Welche Normen gibt es für die Kalibrierung?

-Was bedeutet Kalibrierung, Justierung und Eichung

-Kalibrierungsarten

-Kalibrierungsintervall in GMP-Umgebung

-Wie wichtig ist eine Akkreditierung für ein Kalibrierungservice-Unternehmen?

-Welche Inhalte sollte ein Kalibrierschein in GMP-Umgebung haben?

-Wie wichtig ist die Rückverfolgbarkeit von Kalibrierungsnachweisen?

-Kalibrierungen Inhaus oder Extern?

-Aufbau einer SOP für Kalibrierungen

www.ingramcontent.com/pod-product-compliance
Lightning Source LLC
Chambersburg PA
CBHW070415230526
45471CB00006B/2817